I Spy Space

I Spy Space

Colorful Picture Book

Guessing Game for Little Kids.

HOW TO USE THIS BOOK ?

IN EVERY SECTION, THERE IS ONE LETTER
START GUESSING WHICH ELEMENT.

HOW CAN I CHECK ?

TURN TO THE NEXT PAGE AND
SEE YOUR RESULT

♥ Enjoy ♥

ARE
YOU
READY ?
100

I SPY WITH MY LITTLE EYE SOMETHING BEGINNING WITH :

A

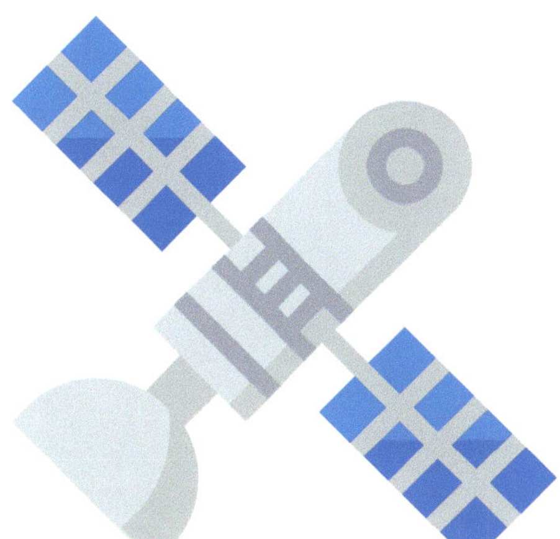

A is for

Asteroid

I SPY WITH MY LITTLE EYE SOMETHING BEGINNING WITH :

B

B is for

Black hole

I SPY WITH MY LITTLE EYE SOMETHING BEGINNING WITH :

C

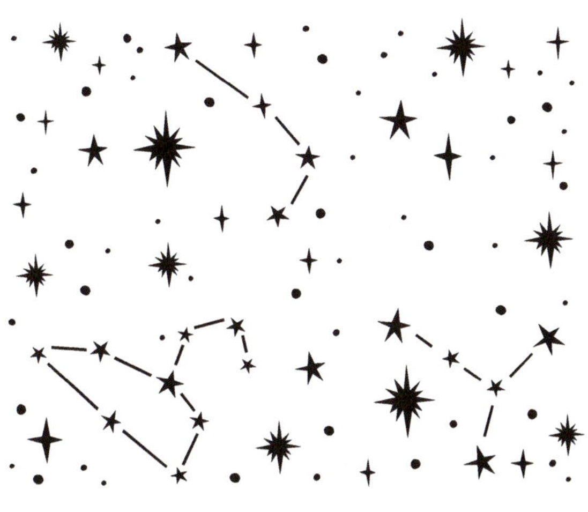

C is for

Constellation

I SPY WITH MY LITTLE EYE SOMETHING BEGINNING WITH :

E

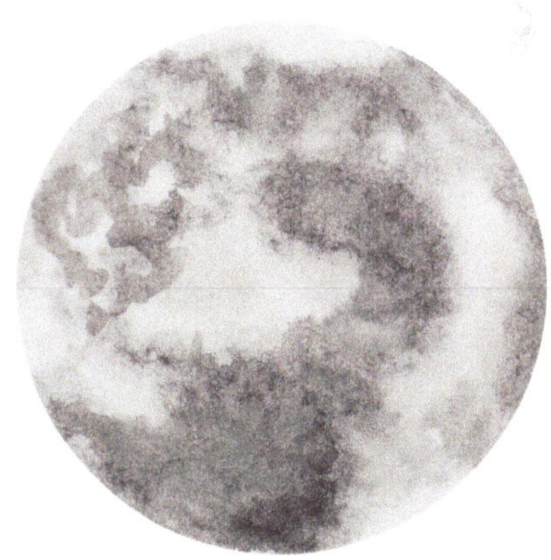

E is for

Eclipse

I SPY WITH MY LITTLE EYE SOMETHING BEGINNING WITH :

F

F is for

Full moon

I SPY WITH MY LITTLE EYE SOMETHING BEGINNING WITH :

G

G is for

Galaxy

I SPY WITH MY LITTLE EYE SOMETHING BEGINNING WITH :

H

H is for

Half moon

I SPY WITH MY LITTLE EYE SOMETHING BEGINNING WITH :

J

J is for

Jupiter

I SPY WITH MY LITTLE EYE
SOMETHING BEGINNING WITH :

M

M is for

Meteor

I SPY WITH MY LITTLE EYE SOMETHING BEGINNING WITH :

N

N is for

Nebula

I SPY WITH MY LITTLE EYE SOMETHING BEGINNING WITH :

O is for

Orbit

I SPY WITH MY LITTLE EYE SOMETHING BEGINNING WITH :

P

P is for

Planet

I SPY WITH MY LITTLE EYE SOMETHING BEGINNING WITH :

Q

Q is for

Quarter moon

I SPY WITH MY LITTLE EYE SOMETHING BEGINNING WITH :

R

R is for

Rocket

I SPY WITH MY LITTLE EYE SOMETHING BEGINNING WITH :

S is for

Solar System

I SPY WITH MY LITTLE EYE SOMETHING BEGINNING WITH :

T

T is for

Telescope

I SPY WITH MY LITTLE EYE SOMETHING BEGINNING WITH :

U

U is for

Universe

I SPY WITH MY LITTLE EYE SOMETHING BEGINNING WITH :

V

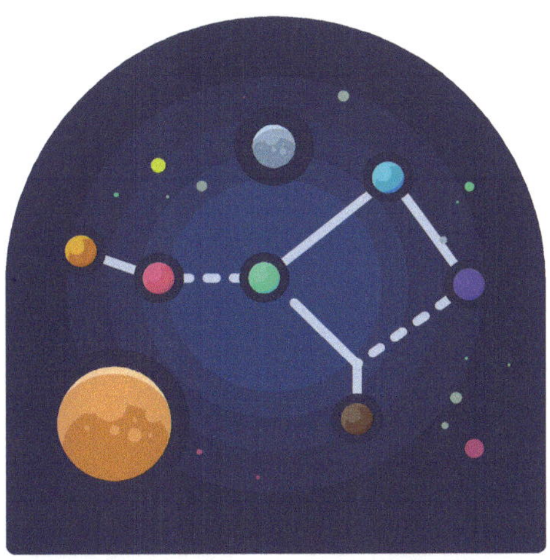

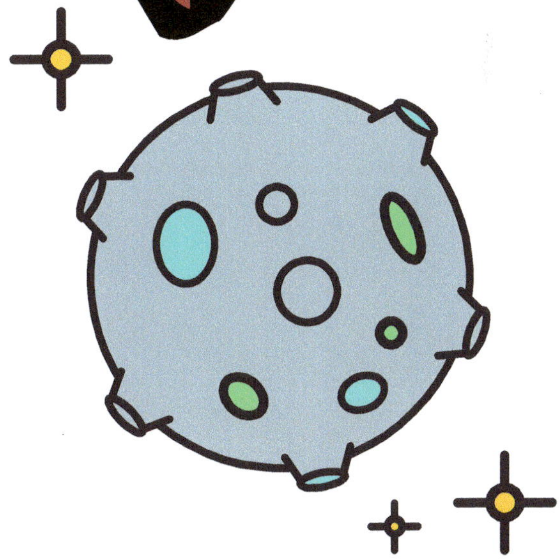

V is for

Venus

www.ingramcontent.com/pod-product-compliance
Lightning Source LLC
Chambersburg PA
CBHW040417220526
45473CB00004B/1272
* 9 7 9 8 7 4 3 3 0 5 2 9 2 *